TIME-ZONES

FLEUR ADCOCK

Oxford   Auckland

OXFORD UNIVERSITY PRESS

1991

Oxford University Press, Walton Street, Oxford OX2 6DP

Oxford New York Toronto
Delhi Bombay Calcutta Madras Karachi
Petaling Jaya Singapore Hong Kong Tokyo
Nairobi Dar es Salaam Cape Town
Melbourne Auckland

and associated companies in
Berlin Ibadan

Oxford is a trade mark of Oxford University Press

First published in Oxford Poets
as an Oxford University Press paperback 1991

British Library Cataloguing in Publication Data
Adcock, Fleur 1934–
Time zones—(Oxford poets)
1. Title
821
ISBN 0–19–282831–2

Library of Congress Cataloging in Publication Data
Adcock, Fleur.
Time-zones / Fleur Adcock.
p. cm.—(Oxford poets)
I. Title. II. Series. 821'.914—dc20
PR9639.3.A3T56 1991 90–42989
ISBN 0–19–282831–2

Typeset by Wyvern Typesetting Ltd.
Printed in Hong Kong

ACKNOWLEDGEMENTS

Poems from this collection have appeared in: *Ambit*, *Antaeus*, *Antipodes New Writing*, the *Belfast Review*, *Bête Noire*, *Descant* (Canada), *First and Always: Poems for Great Ormond Street Children's Hospital*, the *Guardian*, *Island Magazine* (Tasmania), *Landfall*, the *London Review of Books*, the *New Zealand Listener*, *The Orange Dove of Fiji: Poems for the World Wide Fund for Nature*, *Partisan Review*, *P.N. Review*, *Poetry Book Society Anthology* 1987/88 and 1988/89, *Poetry Review*, *The Rialto*, and the *Times Literary Supplement*.

Several poems have been broadcast by the BBC, the Australian Broadcasting Corporation, and the Broadcasting Corporation of New Zealand.

'Next Door' was commissioned by the BBC for *English by Radio*.

'The Farm' was commissioned by the Druridge Bay Campaign for their anthology *Tide-Lines*.

'Counting' was commissioned by Birthright for the exhibition 'Mother & Child'.

'Meeting the Comet' was published by Bloodaxe Books in their series of Pamphlet Poets.

CONTENTS

COUNTING

You count the fingers first: it's traditional.
(You assume the doctor counted them too,
when he lifted up the slimy surprise
with its long dark pointed head and its father's nose
at 2.13 a.m.—'Look at the clock!'
said Sister: 'Remember the time: 2.13.')

Next day the head's turned pink and round;
the nose is a blob. You fumble under the gown
your mother embroidered with a sprig of daisies,
as she embroidered your own Viyella gowns
when you were a baby. You fish out
curly triangular feet. You count the toes.

'There's just one little thing' says Sister:
'His ears—they don't quite match. One
has an extra whorl in it. No one will notice.'
You notice like mad. You keep on noticing.
Then you hear a rumour: a woman in the next ward
has had a stillbirth. Or was it something worse?

You lie there, bleeding gratefully.
You've won the Nobel Prize, and the VC,
and the State Lottery, and gone to heaven.
Feed-time comes. They bring your bundle—
the right one: it's him all right.
You count his eyelashes: the ideal number.

You take him home. He learns to walk.
From time to time you eye him,
nonchalantly, from each side.
He has an admirable nose.
No one ever notices his ears. No one
ever stands on both sides of him at once.

He grows up. He has beautiful children.

LIBYA

When the Americans were bombing Libya
(that time when it looked as if this was it at last,
the match in the petrol-tank which will flare sooner or later,
and the whole lot was about to go up)

Gregory turned on the television during dinner
and Elizabeth asked the children to be quiet
because this was important, we needed to watch the news—
'It might be the beginning of the end' she said.

Oliver, who was seven, said 'But I'm too young to die!'
Lily, who was five, said 'I don't want to die! I don't!'
Oliver said 'I know! Let's get under the table!'
Lily said 'Yes, let's get under the table!'

So they got under the table, and wriggled around our legs
making the dishes rattle, and we didn't stop them
because we were busy straining to hear the news
and watching the fat bombers filling the screen.

It was a noisy ten minutes, one way and another.
Julia, who was fifteen months, chuckled in her high chair,
banging her spoon for her wonderful brother and sister,
and sang 'Three blind mice, three blind mice'.

WHAT MAY HAPPEN

The worst thing that can happen—
to let the child go;
but you must not say so
or else it may happen.

The stranger looms in the way
holding an olive-twig.
The child's not very big;
he is beginning to cry.

How can you stand by?
A cloud crushes the hill.
Everything stands still.
Everything moves away.

The stranger is still a stranger
but the child is not your child.
Too soon, before he's old,
he may become a stranger.

He is his own child.
He has a way to go.
Others have lived it through:
watch, and turn cold.

3

MY FATHER

When I got up that morning I had no father.
I know that now. I didn't suspect it then.
They drove me through the tangle of Manchester
to the station, and I pointed to a sign:

'Hulme' it said—though all I saw was a rubbled
wasteland, a walled-off dereliction. 'Hulme—
that's where they lived' I said, 'my father's people.
It's nowhere now.' I coughed in the traffic fumes.

Hulme and Medlock. A quarter of a mile
to nowhere, to the names of some nothing streets
beatified in my family history file,
addresses on birth and marriage certificates:

Back Clarence Street, Hulme; King Street (but which one?);
One-in-Four Court, Chorlton-upon-Medlock.
Meanwhile at home on my answering machine
a message from New Zealand: please ring back.

In his day it was factory smoke, not petrol,
that choked the air and wouldn't let him eat
until, the first day out from Liverpool,
sea air and toast unlocked his appetite.

He took up eating then, at the age of ten—
too late to cancel out the malnutrition
of years and generations. A small man,
though a tough one. He'll have needed a small coffin.

I didn't see it; he went to it so suddenly,
too soon, with both his daughters so far away:
a box of ashes in Karori Cemetery,
a waft of smoke in the clean Wellington sky.

Even from here it catches in my throat
as I puzzle over the Manchester street-plan,
checking the index, magnifying the net
of close-meshed streets in M2 and M1.

4

Not all the city's motorways and high-rise.
There must be roads that I can walk along
and know they walked there, even if their houses
have vanished like the cobble-stones—that throng

of Adcocks, Eggingtons, Joynsons, Lamberts, Listers.
I'll go to look for where they were born and bred.
I'll go next month; we'll both go, I and my sister.
We'll tell him about it, when he stops being dead.

CATTLE IN MIST

A postcard from my father's childhood—
the one nobody photographed or painted;
the one we never had, my sister and I.
Such feeble daughters—couldn't milk a cow
(watched it now and then, but no one taught us.)
How could we hold our heads up, having never
pressed them into the warm flank of a beast
and lured the milk down? Hiss, hiss, in a bucket:
routine, that's all. Not ours. That one missed us.

His later childhood, I should say;
not his second childhood—that he evaded
by dying—and his first was Manchester.
But out there in the bush, from the age of ten,
in charge of milking, rounding up the herd,
combing the misty fringes of the forest
(as he would have had to learn not to call it)
at dawn, and again after school, for stragglers;
cursing them; bailing them up; it was no childhood.

A talent-spotting teacher saved him.
The small neat smiling boy (I'm guessing)
evolved into a small neat professor.
He could have spent his life wreathed in cow-breath,
a slave to endlessly refilling udders,
companion of heifers, midwife at their calvings,
judicious pronouncer on milk-yields and mastitis,
survivor of the bull he bipped on the nose
('Tell us again, Daddy!') as it charged him.

All his cattle: I drive them back
into the mist, into the dawn haze
where they can look romantic; where they must
have wandered now for sixty or seventy years.
Off they go, then, tripping over the tree-roots,
pulling up short to lip at a tasty twig,
bumping into each other, stumbling off again
into the bush. He never much liked them.
He'll never need to rustle them back again.

6

TOADS

Let's be clear about this: I love toads.

So when I found our old one dying,
washed into the drain by flood-water
in the night and then—if I can bring myself
to say it—scalded by soapy lather
I myself had let out of the sink,
we suffered it through together.

It was the summer of my father's death.
I saw his spirit in every visiting creature,
in every small thing at risk of harm:
bird, moth, butterfly, beetle,
the black rabbit lolloping along concrete,
lost in suburbia; and our toad.

If we'd seen it once a year that was often,
but the honour of being chosen by it
puffed us up: a toad of our own
trusting us not to hurt it
when we had to lift it out of its den
to let the plumber get at the water-main.

And now this desperate damage: the squat
compactness unhinged, made powerless.
Dark, straight, its legs extended,
flippers paralysed, it lay lengthwise
flabby-skinned across my palm,
cold and stiff as the Devil's penis.

I laid it on soil; the shoulders managed
a few slow twitches, pulled it an inch forward.
But the blowflies knew: they called it dead
and stippled its back with rays of pearly stitching.
Into the leaves with it then, poor toad,
somewhere cool, where I can't watch it.

Perhaps it was very old? Perhaps it was ready?
Small comfort, through ten guilt-ridden days.
And then, one moist midnight, out in the country,
a little shadow shaped like a brown leaf
hopped out of greener leaves and came to me.
Twice I had to lift it from my doorway:

a gently throbbing handful—calm, comely,
its feet tickling my palm like soft bees.

UNDER THE LAWN

It's hard to stay angry with a buttercup
threading through the turf (less and less a lawn
with each jagging rip of the fork or scoop
of the trowel) but a dandelion can

inspire righteous fury: that taproot
drilling down to where it's impossible
ever quite to reach (although if it's cut
through that's merely a minor check) until

clunk: what's this? And it's spade-time. Several hours
later, eleven slabs of paving-stone
(submerged so long ago that the neighbours
who've been on the watch since 1941

'never remember seeing a path there') with,
lying marooned singly on three of them,
an octagonal threepence, a George the Fifth
penny and, vaguely missed from their last home

for fifteen years or so and rusted solid,
Grandpa's scissors, the ones for hairdressing
from his barbering days: plain steel, not plated;
still elegant; the tip of one blade still missing.

WREN SONG

How can I prove to you
that we've got wrens in the garden?

A quick flick of a tail
in or out of the ivy hedge
is all you'll ever see of them;

and anyway, I'm asleep.
Not dreaming, though: I can hear him,
the boss-wren, out there in the summer dawn—

his bubbling sequences,
an octave higher than a blackbird's,
trickling silver seeds into my ears.

I'll get the tape-recorder.
But no, it's in another room,
and I've no blank tapes for it;

and anyway, I'm asleep.
Hard to wake up, after a sultry night
of restless dozing, even for the wren.

I've tracked his piccolo solo
in the light evenings, from hedge to apple tree
to elder, sprints of zippy flight in between.

I've looked him up: 'A rapid
succession of penetrating and jubilant
trills, very loud for so small a bird.'

I'll get the tape-recorder.
I'll find an old cassette to record over.
I'm getting up to fetch it now—

but no, I'm still asleep;
it was a dream, the getting up.
But the wren's no dream. It *is* a wren.

NEXT DOOR

You could have called it the year of their persecution:
some villain robbed her window-boxes of half
her petunias and pansies. She wrote a notice:
'To the person who took my plants. I am disabled;
they cost me much labour to raise from seed.'
Next week, the rest went. Then his number-plates.
(Not the car itself. Who'd want the car? It stank.)
A gale blew in a pane of their front window—
crack: just like that. Why theirs? Why not, for example,
mine? Same gale; same row of elderly houses.

And through it all the cats multiplied fatly—
fatly but scruffily (his weak heart, her illness:
'They need grooming, I know, but they're fat as butter')—
and the fleas hopped, and the smell came through the walls.
How many cats? Two dozen? Forty? Fifty?
We could count the ones outside in the cages (twelve),
but inside? Always a different furry face
at a window; and the kittens—think of the kittens
pullulating like maggots over the chairs!
Someone reported them to the authorities.

Who could have done it? Surely not a neighbour!
'No, not a neighbour! Someone in the Fancy'—
she was certain. 'They've always envied my success.
The neighbours wouldn't . . .' A sunny afternoon.
I aimed my camera at them over the fence,
at their garden table, under the striped umbrella:
'Smile!' And they grinned: his gnome-hat, her witch-hair
in the sun—well out of earshot of the door-bell
and of the Environmental Health Inspector.
You could call it a bad year. But the next was worse.

HELIOPSIS SCABRA

This is the time of year when people die:
August, and these daisy-faced things
blare like small suns on their swaying hedge
of leaves, yellow as terror. Goodbye,

they shout to the summer, and goodbye
to Jim, whose turn it was this morning:
while in another hospital his wife
lies paralysed, with nothing to do but lie

wondering what's being kept from her, and cry—
she can still do that. August in hospital
sweats and is humid. In the garden
grey airs blow moist, but the mean sky

holds on to its water. The earth's coke-dry;
the yellow daisies goggle, but other plants
less greedily rooted are at risk.
The sky surges and sulks. It will let them die.

HOUSE-MARTINS

Mud in their beaks, the house-martins are happy . . .
That's anthropomorphism. Start again:

mud being plentiful because last night
it rained, after a month of drought,
the house-martins are able to build their nests.

They flitter under the eaves, white flashes
on their backs telling what they are:
house-martins. Not necessarily happy.

Below in the mock-Tudor cul-de-sac
two kids on skateboards and a smaller girl
with a tricycle are sketching their own circles—

being themselves, being children:
vaguely aware, perhaps, of the house-martins,
and another bird singing, and a scent of hedge.

Anthropomorphism tiptoes away:
of human children it's permissible
to say they're happy—if indeed they are.

It's no use asking them; they wouldn't know.
They may be bored, or in a sulk,
or worried (it doesn't show; and they look healthy).

Ask them in fifty years or so,
if they're still somewhere. Arrange to present them with
(assuming all these things can still be assembled)

a blackbird's song, the honeyed reek of privet,
and a flock of house-martins, wheeling and scrambling
about a group of fake-half-timbered semis.

Call it a Theme Park, if you like:
'Suburban childhood, late 1980s'
(or 70s, or 50s—it's hardly changed).

Ask them 'Were you happy in Shakespeare Close?'
and watch them gulp, sick with nostalgia for it.

WILDLIFE

A wall of snuffling snouts in close-up,
ten coloured, two in black and white,
each in its frame; all magnified,
some more than others. Voles, are they?
Shrews? Water-rats? Whiskers waggling,
they peep from under twelve tree-roots
and vanish. Next, a dozen barn-owls,
pale masks, almost filling the dark screens.
Cut; and now two dozen hedgehogs
come trotting forward in headlong pairs:
they'll fall right out on the floor among the
cookers and vacuum-cleaners unless
the camera—just in time—draws back.
Here they come again, in their various
sizes, on their various grass:
olive, emerald, acid, bluish,
dun-tinged, or monochrome. The tones
are best, perhaps, on the 22-inch
ITT Squareline: more natural
than the Philips—unless you find them too
muted, in which case the Sony
might do. Now here are the owls again.

Meanwhile at the Conference Centre
three fire-engines have screamed up. Not,
for once, a student smoking in a bedroom:
this time a cloud of thunderflies
has chosen to swarm on the pearly-pink
just-warm globe of a smoke-detector.

TURNIP-HEADS

Here are the ploughed fields of Middle England;
and here are the scarecrows, flapping polythene arms
over what still, for the moment, looks like England:
bare trees, earth-colours, even a hedge or two.

The scarecrows' coats are fertilizer bags;
their heads (it's hard to see from the swift windows
of the Intercity) are probably 5-litre
containers for some chemical or other.

And what are the scarecrows guarding? Fields of rape?
Plenty of that in Middle England; also
pillage, and certain other medieval
institutions—some things haven't changed,

now that the men of straw are men of plastic.
They wave their rags in fitful semaphore,
in the March wind; our train blurs past them.
Whatever their message was, we seem to have missed it.

THE BATTERER

What can I have done to earn
the Batterer striding here beside me,
checking up with his blue-china
sidelong eyes that I've not been bad—

not glanced across the street, forgetting
to concentrate on what he's saying;
not looked happy without permission,
or used the wrong form of his name?

How did he get here, out of the past,
with his bulging veins and stringy tendons,
fists clenched, jaw gritted,
about to burst with babble and rage?

Did I elect him? Did I fall
asleep and vote him in again?
Yes, that'll be what he is: a nightmare;
but someone else's now, not mine.

ROLES

Emily Bronte's cleaning the car:
water sloshes over her old trainers
as she scrubs frail blood-shapes from the windscreen
and swirls the hose-jet across the roof.
When it's done she'll go to the supermarket;
then, if she has to, face her desk.

I'm striding on the moor in my hard shoes,
a shawl over my worsted bodice,
the hem of my skirt scooping dew from the grass
as I pant up towards the breathless heights.
I'll sit on a rock I know and write a poem.
It may not come out as I intend.

HAPPINESS

Too jellied, viscous, floating a condition
to inspire more action than a sigh—
like being supported on warm porridge

gazing at this: may-blossom, bluebells, robin,
the tennis-players through the trees,
the trotting magpie (not good news, but handsome)

asking the tree-stump next to where I'm sitting
'Were you a rowan last time? No?
That's what the seedling wedged in your roots is planning.'

COUPLING

On the wall above the bedside lamp
a large crane-fly is jump-starting
a smaller crane-fly—or vice versa.
They do it tail to tail, like Volkswagens:
their engines must be in their rears.

It looks easy enough. Let's try it.

THE GREENHOUSE EFFECT

As if the week had begun anew—
and certainly something has:
this fizzing light on the harbour, these
radiant bars and beams and planes
slashed through flaps and swags of sunny vapour.
Aerial water, submarine light:
Wellington's gone Wordsworthian again.
He'd have admired it—
admired but not approved, if he'd heard
about fossil fuels, and aerosols,
and what we've done to the ozone layer,
or read in last night's *Evening Post*
that 'November ended the warmest spring
since meteorological records began'.
Not that it wasn't wet:
moisture's a part of it.

As for this morning (Friday),
men in shorts raking the beach
have constructed little cairns of evidence:
driftwood, paper, plastic cups.
A seagull's gutting a bin.
The rain was more recent than I thought:
I'm sitting on a wet bench.
Just for now, I can live with it.

THE LAST MOA

Somewhere in the bush, the last moa
is not still lingering in some hidden valley.
She is not stretching her swanlike neck
(but longer, more massive than any swan's)
for a high cluster of miro berries,
or grubbing up fern roots with her beak.

Alice McKenzie didn't see her
among the sandhills at Martin's Bay
in 1880—a large blue bird
as tall as herself, which turned and chased her.
Moas were taller than seven-year-old
pioneer children; moas weren't blue.

Twenty or thirty distinct species—
all of them, even the small bush moa,
taller than Alice—and none of their bones
carbon-dated to less than five centuries.
The sad, affronted mummified head
in the museum is as old as a Pharoah.

Not the last moa, that; but neither
was Alice's harshly grunting pursuer.
Possibly Alice met a takahe,
the extinct bird that rose from extinction
in 1948, near Te Anau.
No late reprieve, though, for the moa.

Her thigh-bones, longer than a giraffe's,
are lying steeped in a swamp, or smashed
in a midden, with her unstrung vertebrae.
Our predecessors hunted and ate her,
gobbled her up: as we'd have done
in their place; as we're gobbling the world.

CREOSOTE

What is it, what is it? Quick: that whiff,
that black smell—black that's really brown,
sharp that's really oily and yet rough,

a tang of splinters burning the tongue,
almost as drunkening as hot tar
or cowshit, a wonderful ringing pong.

It's fence-posts, timber yards, the woodshed;
it bundles you into the Baby Austin
and rushes you back to early childhood.

It's Uncle's farm; it's the outside dunny;
it's flies and heat; or it's boats and rope
and the salt-cracked slipway down from the jetty.

It's brushes oozing with sloshy stain;
it's a tin at the back of the shed: open it,
snort it! You can't: the lid's stuck on.

CENTRAL TIME

'The time is nearly one o'clock,
or half past twelve in Adelaide' —
where the accents aren't quite so . . . Australian
as in the other states, the ones
that were settled (not their fault, of course)
by convicts. We had Systematic
Colonization, and Colonel Light,
and the City of Adelaide Plan. We have the Park Lands.

It's time for the news at 1.30 —
one o'clock Central Time in Adelaide.

It's early days in Hobart Town,
and Maggie May has been transported
(not such fun as it sounds, poor lass)
to toil upon Van Diemen's cruel shore.
It's 1830 or thereabouts
(1800 in Adelaide?
No, no, this is going too far —
as she might have said herself at the time).

The time is three o'clock, etc.
The time is passing.
You're tuned to ABC Radio.
We'll be bringing you that programme shortly.

It's five o'clock in Adelaide
and Maggie May has found her way
to a massage parlour in Gouger Street.
The Red Light Zone (as we don't call it)
extends from the West Park Lands to Light Square
(named for the Colonel, not the Zone).
The Colonel's in two minds about it;
his fine Eurasian face is troubled.

The Colonel's an anomaly.
There are plenty of those in Adelaide.

Meanwhile, back in Van Diemen's Land,
a butcher bird sings coloratura
in the courtyard of the Richmond Gaol
as tourists file through with their cameras,
wondering how to photograph
a Dark Cell for solitary
from the inside, with the door shut.
Look, they had them for women too!

It's half past eight in Adelaide
and 4 a.m. in Liverpool.
Maggie May wants to ring Lime Street.
You mean they don't have STD?
But I thought this was the New World.

They don't have GMT either;
or BST, as they call it now,
whenever now is.
                    It is now
half past ten in Adelaide,
and in the Park Lands a nasty man
is cutting up a teenage boy
and cramming him into a plastic bag.

In Gouger Street another man,
equally nasty but less wicked,
has taken his wife to a performance
of Wagner at the Opera Theatre
and is strolling with her to their car
past the massage parlour
where something like five hours ago
Maggie May gave him a hand-job.

The Colonel's brooding over his notebooks,
and lying under his stone, and standing
on his plinth on Montefiore Hill.

Maggie May is still on the phone,
arguing with the operator,
trying to get through to Lime Street.
It's the future she wants,
or the past back. Some of it.

You're listening to ABC FM:
12.30 Eastern Standard Time—
twelve midnight in Adelaide.
And now, to take us through the night,
Music to Keep the Days Apart.

THE BREAKFAST PROGRAM

May: autumn. In more or less recognizable
weather, more or less recognizable birds
are greeting the dawn. On 5CL the newsreader
has been allotted (after the lead story
on whether the Treasurer might or still might not
cancel the promised tax-cuts) two minutes
to tell us about whatever it is today—
chemical weapons, radioactive rain,
one of those messy bits of northern gloom
from the places where gloom's made (not here, not here!).
He tells us; then the baby-talking presenter
(curious how some Australian women
never get to sound older than fifteen)
contrives a soothing link: 'Grim news indeed'
she ad-libs cosily. 'Much worse, of course,
if you live in Europe'—writing off a hemisphere.

FROM THE DEMOLITION ZONE

Come, literature, and salve our wounds:
bring dressings, antibiotics, morphine;

bring syringes, oxygen, plasma.
(Saline solution we have already.)

We're injured, but we mustn't say so;
it hurts, but we mustn't tell you where.

Clear-eyed literature, diagnostician,
be our nurse and our paramedic.

Hold your stethoscope to our hearts
and tell us what you hear us murmuring.

Scan us; but do it quietly, like
the quiet seep of our secret bleeding.

When we lie awake in the night
cold and shaking, clenching our teeth,

be the steady hand on our pulse,
the skilful presence checking our symptoms.

You know what we're afraid of saying
in case they hear us. Say it for us.

ON THE WAY TO THE CASTLE

It would be rude to look out of the car windows
at the colourful peasants authentically pursuing
their traditional activities in the timeless landscape
while the editor is talking to us.
He is telling us about the new initiatives
his magazine has adopted as a result
of the Leader's inspiring speech at the last Party Congress.
He is speaking very slowly (as does the Leader,
whom we have seen on our hotel television),
and my eyes are politely fixed on his little moustache:
as long as it keeps moving they will have to stay there;
but when he pauses for the interpreter's turn
my duty is remitted, and I can look out of the windows.
I am not ignoring the interpreter's translation
but she has become our friend: I do not feel compelled
by courtesy to keep my eyes on her lipstick.
What's more, the editor has been reciting his speech
at so measured a pace and with such clarity
that I can understand it in his own language;
and in any case, I have heard it before.
This on-off pattern of switching concentration
between the editor's moustache and the sights we are passing
gives me a patchy impression of the local agriculture.
Hordes of head-scarved and dark-capped figures
move through fields of this and that, carrying implements,
or bending and stretching, or loading things on to carts.
I missed most of a village, during the bit about the print-run,
but the translation granted me a roadful of quaint sheep.
Now the peasants are bent over what looks like bare earth
with occasional clusters of dry vegetation.
It is a potato-field; they are grubbing for potatoes.
There are dozens of them—of peasants, that is:
the potatoes themselves are not actually visible.
As a spectacle, this is not notably picturesque,
but I should like to examine it for a little longer.
The sky has turned black; it is beginning to rain.
The editor has thought of something else he wishes to tell us
about the magazine's history.
Once again, eyes back to his official moustache
(under which his unofficial mouth looks vulnerable).

The editor is a kind man.
He is taking us on an interesting excursion,
in an expensive taxi, during his busy working day.
It has all been carefully planned for our pleasure.
Quite possibly he wants to shield us from the fact
that this rain is weeks or months too late;
that the harvest is variously scorched, parched and withered;
that the potatoes for which the peasants are fossicking
have the size and the consistency of bullets.

ROMANIA

Suddenly it's gone public; it rushed out
into the light like a train out of a tunnel.
People I've met are faces in the government,
shouting on television, looking older.

The country sizzles with freedom. The air-waves
tingle. The telephone lines are all jammed.
I can't get through to my friends. Are they safe? They're safe,
but I need to hear it from them. Instead

I'll play the secret tape I made in the orchard
two years ago, at Ciorogîla.
We're talking in two languages, mine and theirs,
laughing, interrupting each other;

the geese in the peasants' yard next door
are barking like dogs; the children are squawking,
chasing each other, picking fruit;
the little boy brings me a flower and a carrot.

We're drinking must—blood-pink, frothy—
and a drop of unofficial ţuica:
'What do the peasants drink in your country?—
Oh, I forgot, you don't have peasants.'

It's dusk. The crickets have started up:
Zing-zing, zing-zing, like telephones
over the static. Did it really happen?
Is it possible? 'Da, da!' say the geese.

December 1989

CAUSES

THE FARM

(*In memory of Fiona Lodge*)

Fiona's parents need her today—
they're old; one's ill, and slipping away—
but Fiona won't be by the bed:
                    she's dead.
She went for a working holiday
years ago, on a farm that lay
just down the coast from St Bee's Head
in Cumbria, next to—need I say?
                    A name to dread.
She was always very fond of the farm
with its rough, authentic rural charm,
and the fields she tramped, and the lambs she fed
                    with youthful pride.
Her family saw no cause for alarm—
how could it do her any harm
working there in the countryside?
It would help to build her up, they said.
But it secretly broke her down instead,
                    until she died.
There was a leak, if you recall,
at Windscale in the fifties. No?
Well, it was thirty years ago;
                    but these things are slow.
And no matter what the authorities said
about there being no risk at all
from the installations at Calder Hall,
buckets of radiation spread,
                    and people are dead.
That farm became a hazardous place—
though to look at it you wouldn't know;
but cancers can take years to grow
(or leukaemia, in Fiona's case),
and as often as not they win the race,
                    however slow.

31

Before long most of us will know
people who've died in a similar way.
We're not aware of it today,
                    and nor are they,
but another twenty years or so
will sort out who are the ones to go.
We'll be able to mark them on a chart,
a retrospective map to show
where the source of their destruction lay.
                    That's the easy part.
But where's the next lot going to start?
At Windscale, Hinkley Point, Dounreay,
Dungeness, Sizewell, Druridge Bay?
                    Who can say?

ALUMINIUM

Ting-ting! 'What's in your pocket, sir?'
Ping! Metal. Not coins or keys:
Sterotabs for the foreign water,
armour against one kind of disease.

'Aluminium: that's what they are—
they set the machine off.' That's it, then:
out of the frying-pan into the fire;
here's awful Alzheimer's looming again.

There wasn't much point in throwing away
your aluminium pots and kettle
if whenever you go on holiday
your drinking water's full of that metal.

Which will you swallow: bacteria soup,
or a clanking cocktail of sinister granules
that'll rust your mental circuitry up
and knot your brain-cells into tangles?

Don't bother to choose. You can't abjure it,
the use of this stuff to 'purify'.
At home the Water Board's fallen for it:
don't be surprised to see a ring of sky,

grey and canny as a metal detector,
to hear, amidst an aerial hum,
tintinnabulations over the reservoir
warning you of dementia to come.

A HYMN TO FRIENDSHIP

Somehow we manage it: to like our friends,
to tolerate not only their little ways
but their huge neuroses, their monumental oddness:
'Oh well', we smile, 'it's one of his funny days.'

Families, of course, are traditionally awful:
embarrassing parents, ghastly brothers, mad aunts
provide a useful training-ground to prepare us
for the pseudo-relations we acquire by chance.

Why them, though? Why not the woman in the library
(grey hair, big mouth) who reminds us so of J?
Or the one on Budgen's delicatessen counter
(shy smile, big nose) who strongly resembles K?

—Just as the stout, untidy gent on the train
reading 'The Mail on Sunday' through pebble specs
could, with somewhat sparser hair and a change
of reading-matter, be our good friend X.

True, he isn't; they aren't; but why does it matter?
Wouldn't they do as well as the friends we made
in the casual past, by being at school with them,
or living next door, or learning the same trade?

Well, no, they wouldn't. Imagine sharing a tent
with one of these look-alikes, and finding she snored:
no go. Or listening for days on end while she dithered
about her appalling marriage: we'd be bored.

Do we feel at all inclined to lend them money?
Or travel across a continent to stay
for a weekend with them? Or see them through an abortion,
a divorce, a gruelling court-case? No way.

Let one of these impostors desert his wife
for a twenty-year-old, then rave all night about
her sensitivity and her gleaming thighs,
while guzzling all our whisky: we'd boot him out.

And as for us, could we ring them up at midnight
when our man walked out on us, or our roof fell in?
Would they offer to pay our fare across the Atlantic
to visit them? The chances are pretty thin.

Would they forgive our not admiring their novel,
or saying we couldn't really take to their child,
or confessing that years ago we went to bed
with their husband? No, they wouldn't: they'd go wild.

Some things kindly strangers will put up with,
but we need to know exactly what they are:
it's OK to break a glass, if we replace it,
but we mustn't let our kids be sick in their car.

Safer to stick with people who remember
how we ourselves, when we and they were nineteen,
threw up towards the end of a student party
on ethyl alcohol punch and methedrine.

In some ways we've improved since then. In others
(we glance at the heavy jowls and thinning hair,
hoping we're slightly better preserved than they are)
at least it's a deterioration we share.

It can't be true to say that we chose our friends,
or surely we'd have gone for a different lot,
while they, confronted with us, might well have decided
that since it was up to them they'd rather not.

But something keeps us hooked, now we're together,
a link we're not so daft as to disparage—
nearly as strong as blood-relationship
and far more permanent, thank God, than marriage.

35

SMOKERS FOR CELIBACY

Some of us are a little tired of hearing that cigarettes kill.
We'd like to warn you about another way of making yourself
   ill:

we suggest that in view of AIDS, herpes, chlamydia, cystitis
   and NSU,
not to mention genital warts and cervical cancer and the
   proven connection between the two,

if you want to avoid turning into physical wrecks
what you should give up is not smoking but sex.

We're sorry if you're upset,
but think of the grisly things you might otherwise get.

We can't see much point in avoiding emphysema at sixty-five
if that's an age at which you have conspicuously failed to
   arrive;

and as for cancer, it is a depressing fact
that at least for women this disease is more likely to occur in
   the reproductive tract.

We could name friends of ours who died that way, if you
   insist,
but we feel sure you can each provide your own list.

You'll notice we didn't mention syphilis and gonorrhoea;
well, we have now, so don't get the idea

that just because of antibiotics quaint old clap and pox
are not still being generously spread around by men's cocks.

Some of us aren't too keen on the thought of micro-organisms
   travelling up into our brain
and giving us General Paralysis of the Insane.

We're opting out of one-night stands;
we'd rather have a cigarette in our hands.

If it's a choice between two objects of cylindrical shape
we go for the one that is seldom if ever guilty of rape.

Cigarettes just lie there quietly in their packs
waiting until you call on one of them to help you relax.

They aren't moody; they don't go in for sexual harassment
    and threats,
or worry about their performance as compared with that of
    other cigarettes,

nor do they keep you awake all night telling you the story of
    their life,
beginning with their mother and going on until morning
    about their first wife.

Above all, the residues they leave in your system are
    thoroughly sterilized and clean,
which is more than can be said for the products of the human
    machine.

Altogether, we've come to the conclusion that sex is a drag.
Just give us a fag.

MRS FRASER'S FRENZY

Songs for Music

1

My name is Eliza Fraser.
I belong to some savages.
My job is to feed the baby
they have hung on my shoulder.

Its mother is lying sick
with no milk in her breasts,
and my own baby died:
it was born after the shipwreck.

It was born under water
in the ship's leaky longboat.
Three days I helped to bail,
then gave birth in the scuppers.

My poor James, the captain,
was crippled with thirst and sickness.
The men were all useless,
and no woman to call on.

I believe the First Mate,
Mr Brown, treated me kindly;
he consigned my dead infant
to its watery fate.

But now I have been given
a black child to suckle.
I have been made a wet-nurse,
a slave to savage women.

They taunt me and beat me.
They make me grub for lily-roots
and climb trees for honey.
They poke burning sticks at me.

They have rubbed me all over
with charcoal and lizard-grease
to protect me from sunburn.
It is my only cover.

I am as black as they are
and almost as naked,
with stringy vines for a loincloth
and feathers stuck in my hair.

They are trying to change me
into one of themselves.
My name is Eliza Fraser.
I pray God to save me.

Their men took my husband—
they dragged him into the forest—
but I still have my wedding-ring
concealed in my waistband.

My name is Eliza Fraser.
My home is in Stromness.
I have left my three children
in the care of the minister.

I am a strong woman.
My language is English.
My name is Eliza Fraser
and my age thirty-seven.

2

The ghosts came from the sea, the white ghosts.
One of them was a she-ghost, a white woman.
We took her to the camp, the white she-ghost.
She was white all over, white like the ancestors,
white like the bodies of dead people
when you scorch them in the fire and strip off the skin.
She was a ghost, but we don't know whose.
We asked her 'Whose ghost are you?
Which ancestor has come back to us?'
She wouldn't say. She had forgotten our language.
She talked in a babble like the babble of birds,
that ghost from the sea, that white she-ghost.
She was covered with woven skins, but we stripped her;
she had hairs on her body, but we plucked them out;

we tried to make her look like a person.
She was stupid, though. She wouldn't learn.
We talked to her and she didn't listen.
We told her to go out and collect food, to dig for roots.
We told her to climb trees, to look for honey.
She couldn't, not even when we beat her.
She seems to have forgotten everything,
that ghost from the sea, that white woman.
We send her out for food every day
and she brings back a few bits, not enough for a child.
We have to throw her scraps, or she would starve.
All she is fit for is to suckle a baby,
that ancestor woman, that white ghost.
We have put her among the children until she learns.

3

I am a poor widow.
I do not own a farthing—
bereft in a shipwreck
of all but my wedding-ring.

    You are a liar, Mrs Fraser.
    You own two trunks of finery
    and £400 subscribed
    by the citizens of Sydney.

I am a poor widow.
My fatherless children
are alone up in Orkney
while I beg for money.

The Lord Mayor of Liverpool,
the Lord Mayor of London,
the Colonial Secretary:
they will none of them help me.

    You are a liar, Mrs Fraser.
    You are not even Mrs Fraser.
    You have another husband now—
    you married Captain Greene in Sydney.

I am a poor widow,
the victim of cannibals.
They killed my dear husband
on the shores of New Holland.

They skinned him and baked him;
they cut up his body
and gorged on his flesh
in their villainous gluttony.

Their hair is bright blue,
those abominable monsters;
it grows in blue tufts
on the tips of their shoulders . . .

    You are a liar, Mrs Fraser.
    Your sad ordeals have quite unhinged you.
    You were a decent woman once,
    prickly with virtue. What has changed you?
    Tell us the truth, the truth, the truth!
    What really happened that deranged you?

4
Not easy to love Mrs Fraser.
Captain Fraser managed it, in his time—
hobbling on her arm, clutching his ulcer,
falling back to relieve his griping bowels;
and hauling timber, a slave to black masters:
'Eliza, wilt thou help me with this tree?—
Because thou art now stronger than me.'
But they speared him, and she fainted, just that once.

Her children had to love her from a distance—
from Orkney to the far Antipodes,
or wherever she'd sailed off to with their father,
cosseting him with jellies for his gut:
'I have received a letter from dear Mamma.
I am looking for her daily at Stromness.'
Daily they had no sight of Mrs Fraser—
who had secretly turned into Mrs Greene.

And Captain Greene? Did he contrive to love her?
He never saw her as her rescuers did:
'Perfectly black, and crippled from her sufferings,
a mere skeleton, legs a mass of sores.'
He saw a widow, famous, with some money.
He saw the chance of more. He saw, perhaps,
a strangeness in her, gone beyond the strangeness
of anything he'd met on the seven seas.

5

I am not mad. I sit in my booth
on show for sixpence: 'Only survivor'
(which is a lie) 'of the Stirling Castle
wrecked off New Holland' (which is the truth),

embroidering facts. There is no need
to exaggerate (but I do), to sit
showing my scars to gawping London.
I do it for money. This is not greed:

I am not greedy. I am not mad.
I have a husband. I am cared for.
But I wake in the nights howling, naked,
alone, and starving. All that I had

I lost once—all the silken stuff
of civilization: clothes, possessions,
decency, liberty, my name;
and now I can never get enough

to replace it. There can never be
enough of anything in the world,
money or goods, to keep me warm
and fed and clothed and safe and free.

MEETING THE COMET

BEFORE

1

She'll never be able to play the piano—
well, not properly. She'll never be able
to play the recorder, even, at school,
when she goes: it has so many little holes . . .

We'll have her taught the violin.
Lucky her left hand's the one with four
fingers, one for each string. A thumb
and a fleshy fork are enough to hold a bow.

2

Before the calculator—the electronic one—
there were beads to count on; there was the abacus
to tell a tally or compute a score;
or there were your fingers, if you had enough.

The base was decimal: there had to be
a total of ten digits, in two sets—
a bunch of five, another bunch of five.
If they didn't match, your computations went haywire.

3

On the left hand, four	and a thumb.
On the right, a thumb	and just two.
Proper fingers, true,	fitted out
in the standard way;	but not four.

Baby-plump, the wrist	on the left.
On the right, the arm	narrows down
to a slender stem	and a palm
like a little tube	of soft bones.

4

Leafy lanes and *rus in urbe* were the thing
for a sheltered childhood (not that it was for long,
but parents try): the elm trees lingering
behind the coach factory; the tense monotonous song

of collared doves; the acres of bare floor
for learning to gallop on in the first size
of Start-Rite shoes; the peacock glass in the front door;
and the swift refocusing lurch of the new baby-sitter's eyes.

5

The Duke of Edinburgh stance: how cute
in a five-year-old! She doesn't do it much
when you're behind her; then it's hands in armpits
or pockets. School, of course, would like to teach

that well-adjusted children don't need pockets
except for their normal purposes, to hold
hankies or bus-tickets. She'll not quite learn
what she's not quite specifically taught.

6

Perhaps I don't exist. Perhaps
I didn't exist till I thought that;
then God invented me and made me
the age I am now (nearly eight);

perhaps I was someone else before,
and he suddenly swapped us round, and said
'You can be the girl with two fingers
and she can be you for a change, instead.'

7

'Give us your hand—it's a bit muddy here,
you'll slip.' But he's on her wrong side: her right's
wrong. She tries to circumnavigate him
('Watch it!' he says), to offer him her left—

and slips. It comes out. 'There!' she says. 'You see!'
'Is that all? Fucking hell', he says, 'that's nothing;
don't worry about it, love. My Auntie May
lost a whole arm in a crash. Is it hereditary?'

8

'Some tiny bud that should have split into four
didn't, we don't know why' was all they could offer.
Research, as usual, lags. But suddenly, this:
'A long-term study has found a positive link

between birth defects and exposure to pesticides
in the first twelve weeks of pregnancy . . . the baby's
neural crest . . . mothers who had been present when
aerosol insecticides . . .' Now they tell us.

TRAVELLING

9 So Far

She has not got multiple sclerosis.
She has not got motorneuron disease,
or muscular dystrophy, or Down's Syndrome,
or a cleft palate, or a hole in the heart.

Her sight and hearing seem to be sound.
She has not been damaged by malnutrition,
or tuberculosis, or diabetes.
She has not got (probably not got) cancer.

10 Passport

Date of birth and all that stuff: straightforward;
likewise, now that she's stopped growing, height.
But ah, 'Distinguishing marks': how can she smuggle
so glaring a distinction out of sight?

The Passport Office proves, in one of its human
incarnations, capable of tact:
a form of words emerges that fades down
her rare statistic to a lustreless fact.

11 Stars

She's seeing stars—Orion steady on her left
like a lit-up kite (she has a window-seat),
and her whole small frame of sky strung out
with Christmas-tree lights. But what's all that

behind them? Spilt sugar? Spangled faults
in the plane's window? A dust of glittering points
like the sparkle-stuff her mother wouldn't let her
wear on her eyes to the third-form party.

12 Halfway

Does less mean more? She's felt more nearly naked
in duffel-coat and boots and scarf
with nothing showing but a face and her bare
fingers (except, of course, for the times

in fur gloves—mittens—look, no hands!)
than here on a beach in a bikini:
flesh all over. Look at my legs, my
back, my front. Shall I take off my top?

13 At the Airport

Shoulders like horses' bums; an upper arm
dressed in a wobbling watermelon of flesh
and a frilly muu-muu sleeve; red puckered necks
above the bougainvillaea and sunsets

and straining buttons of Hawaiian shirts;
bellies, bald heads, a wilting grey moustache
beneath a hat proclaiming 'One Old Poop'.
The tour guide rounds them up: his travelling freak-show.

14 Comet

'There will be twenty telescopes in the crater
of Mount Albert.' White-coated figures man them,
marshalling queues in darkness: not the Klan
but the Lions raising funds for charity.

$2 a look. No lights—not even torches;
no smoking (bad for the optics); no moon
above the tree-fringed walls of this grassy dip.
Nothing up there but stars. And it, of course.

15 Halley Party

A glow-worm in a Marmite jar
like the one her mother brought her once:
'I dreamt you woke me in the night and showed me
a glow-worm in a Marmite jar.'

So these wee kids in dressing-gowns
will remember being woken up
for honey sandwiches and cocoa
and a little light in a ring of glass.

16 Orbit

'It's not like anything else, with its stumpy tail:
just a fuzz, really, until you get up close —
but of course you can't. With binoculars, I meant,
or a telescope. Actually the tail's fading.'

Higher than Scorpius now, higher than the Pointers,
high as the mid-heaven, she's tracked it nightly,
changing. 'I'm not the only one, but I'm once
in a lifetime.' As for close, that's something else.

AFTER

17

Landing at Gatwick on a grey Sunday when
the baggage handlers seem to be on strike as
they were at the airport before last (but no,
it's merely Britain being its old self) she's

her old self—a self consisting also of
more hand-luggage than she'd thought she was allowed
plus her at last reclaimed suitcase: all of which,
however she may dispose them, hurt her hand.

18

Rise above it! Swallow a chemical:
chuck down whisky, Valium, speed,
Mogadon, caffeine; bomb it or drown it.
But wait! If chemicals did the deed

pandering to their ways compounds
the offence. Resist: you know they lead
to trouble. Find another obsession.
Face a healthier form of need.

51

19

Saving the world is the only valid cause.
Now that she knows it's round it seems smaller,
more vulnerable (as well as bigger, looser,
a baggy bundle of dangerous contradictions).

There's room for such concerns in student life,
if you stretch it. So: Link hands around the world
for peace! Thumbs down to Star Wars! Hands off
the environment! Two fingers to the Bomb!

20

'Of course you'd have a natural sympathy . . .
I always thought it was quite sweet, your little hand,
when we were kids; but we don't want other kids
walking around the world with worse things . . .

I'm not upsetting you, am I?' No, she's not,
this warm voice from the past, this candid face.
'Right. See you tomorrow. The coach leaves at 8.
Oh, and we've got a wonderful furious banner.'

21

The fountain in her heart informs her
she needn't try to sleep tonight—
rush, gush: the sleep-extinguisher
frothing in her chest like a dishwasher.

She sits at the window with a blanket
to track the turning stars. A comet
might add some point. The moon ignores her;
but dawn may come. She'd settle for that.

22

There was a young woman who fell
for someone she knew rather well—
a friend from her school: confirming the rule
that with these things you never can tell.

The person she'd thought a fixed star—
stuck on rails like a tram, not a car—
shot off into orbit and seemed a new planet,
and a dazzler, the finest by far.

23

She wants to see what it looks like on
a breast. She puts it on a breast—
not the one she has in mind
but her own: at least it's a rehearsal.

Three weeks later, the first night:
a nipple, darker than hers, framed
in a silky, jointed bifurcation.
There is also dialogue. And applause.

24

And she never did learn to play the violin.
So it will have to be *Musica Mundana*,
'the harmony of the spheres' (coming across a map
of the southern skies cut out of some Auckland paper)

or the other kind: what was it? *Instrumentalis*
and—ah, yes—*Humana*. (Listen: Canopus, Crux,
Carina, Libra, Vela choiring together. She
has glided right off the edge of the star-chart.)

OXFORD POETS

Fleur Adcock
James Berry
Edward Kamau Brathwaite
Joseph Brodsky
Basil Bunting
W. H. Davies
Michael Donaghy
Keith Douglas
D. J. Enright
Roy Fisher
David Gascoyne
Ivor Gurney
David Harsent
Anthony Hecht
Zbigniew Herbert
Thomas Kinsella
Brad Leithauser
Derek Mahon

Medbh McGuckian
Jamie McKendrick
James Merrill
Peter Porter
Craig Raine
Christopher Reid
Stephen Romer
Carole Satyamurti
Peter Scupham
Penelope Shuttle
Louis Simpson
Anne Stevenson
George Szirtes
Grete Tartler
Edward Thomas
Charles Tomlinson
Chris Wallace-Crabbe
Hugo Williams